World of science

ENERGY FROM SUN AND ATOM

BAY BOOKS LONDON & SYDNEY

1980 Published by Bay Books
157–167 Bayswater Road, Rushcutters
Bay NSW 2011 Australia

National Library of Australia
Card Number and ISBN 0 85835 275 3
Design: Sackville Design Group
Printed by Tien Wah Press, Singapore.

ATOMS AND MOLECULES

All substances are made up of atoms, and in many substances the atoms are grouped together in a certain way. A group of atoms combined together is called a *molecule.* In a pure substance, all the atoms or molecules are identical. On the other hand, they are different from the molecules in other substances.

Molecules

In most substances, the molecules contain several atoms. In a water molecule, for example, there are only three atoms: two hydrogen atoms and one oxygen atom. If the molecules of a substance are broken apart, either into individual atoms or smaller groups of atoms, then the substance changes into something else. For example, if an electric charge is passed through a water molecule, it will break down into hydrogen gas, which contains only hydrogen atoms, and oxygen gas, which contains only oxygen atoms.

Until the end of the nineteenth century, scientists were convinced that atoms were the smallest and simplest

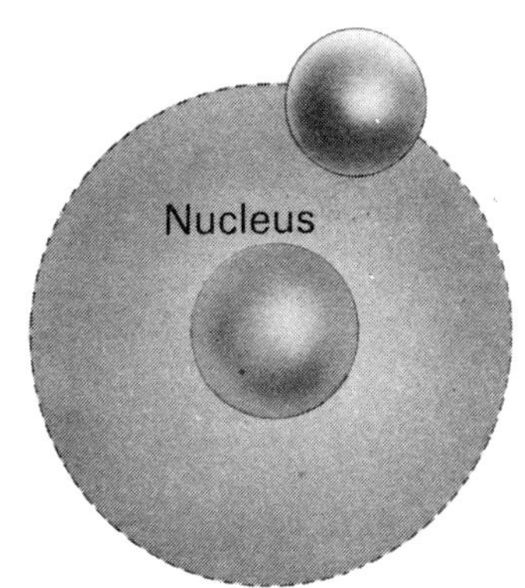

Hydrogen atom

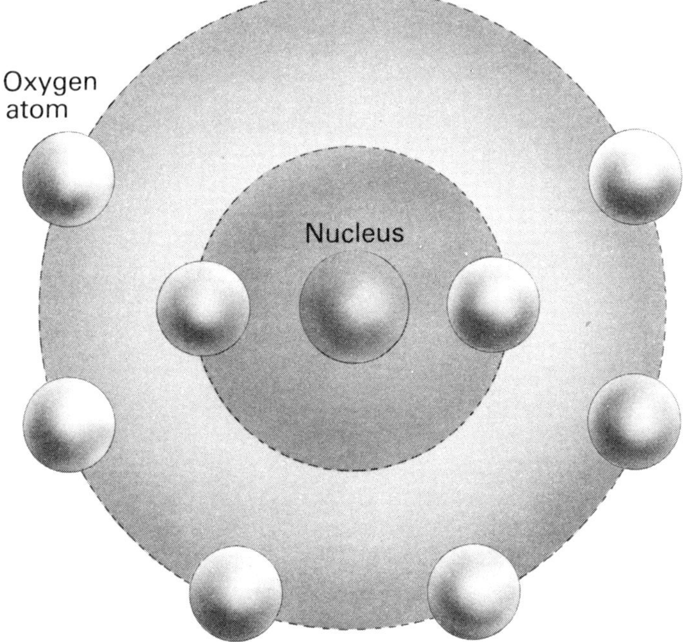

Eight electrons in two shells

Every atom consists of a small, central nucleus which is surrounded by a 'cloud' of one or more electrons, each with a negative electric charge. Thus the electrons mutually repel each other. Atoms are the basic building blocks for every substance and organism in the world around us.

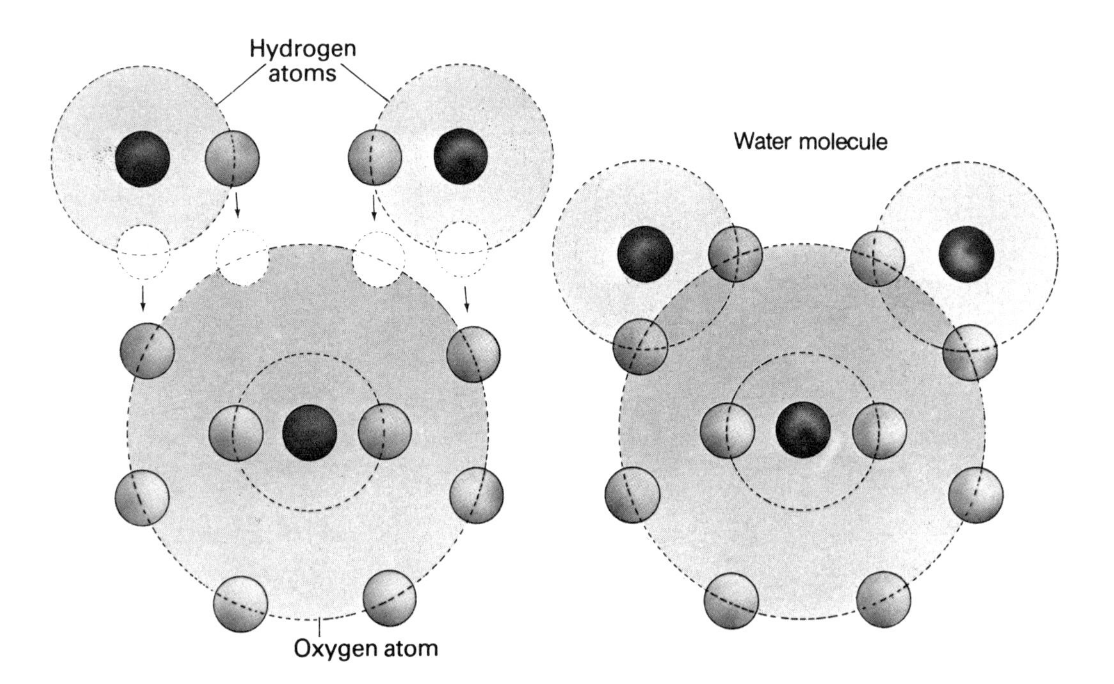

The clouds of electrons of different atoms may sometimes interact to create a chemical bond, and thus a molecule is formed. When an oxygen atom and two hydrogen atoms are close together, some of the electrons are 'shared' between pairs of atoms, which then bond to form a stable molecule of water with the chemical formula H_2O.

particles of matter. At that time, however, smaller particles than atoms were discovered. These were called electrons and they are, in fact, parts of atoms. A number of other very small particles have been discovered and the study of such particles is called *particle physics.*

Parts of the atom

In 1911, Ernest Rutherford discovered the atomic nucleus, enabling scientists to find the basic structure of atoms making up elements such as hydrogen. In 1932 another scientist, James Chadwick, discovered the part of the atom structure we call the *neutron.*

Atoms are now known to consist of *protons,* which are positively charged, *electrons,* which are negatively charged, and *neutrons* which are neutral.

Surprisingly, atoms are almost empty space. Most of the mass of the atom is in its *nucleus,* which is tiny in proportion to the whole. Around the nucleus or centre move one or more of the smaller particles called electrons. The outer boundary of an atom is the path made by its outermost electrons and the nucleus is only a ten-thousandth of the diameter of the atom in size.

The nucleus of most atoms consists of protons and neutrons. The protons have a positive electrical charge and the neutrons have no charge. But the electrons moving around the nucleus have an electrical charge equal to, and opposite to, the charge of the proton. In all atoms, the positive and negative charges balance each other so that the atoms are electrically neutral.

The simplest atom is the hydrogen atom. It consists of a nucleus of one proton around which one electron moves. It has no neutrons. The nuclei of other elements all contain increasing numbers of protons, neutrons and electrons. Helium, the next element, has an atom with a nucleus of two protons and two neutrons and two electrons moving around the nucleus. One of the most complex atoms is the uranium atom. It has a nucleus of 92 protons and 146 neutrons, with 92 electrons surrounding it.

Ionised atoms

Although an atom is mostly empty space, one atom cannot move inside another. The negative charges of the electrons moving around the nucleus mean that the out-

The lightest and most abundant element in the universe, hydrogen is present in all living things. This colourless, odourless and tasteless gas forms a large part of stars, clouds of dust and nebulae in space, as in the Trifid nebula in the constellation of Sagittarius shown here.

side of each atom has a negative charge. As similar charges repel each other, atoms remain separate from each other. But if an atom loses one of its electrons or gains an extra one from somewhere else, then the charge of the atom will change. This can easily happen and when it does the atoms that have acquired a positive charge will now be attracted to atoms with a negative charge. These atoms are now *ionised* and are called *ions.* They can combine to form other compounds, or under certain conditions they can produce energy.

ENERGY FROM ATOMS

Even though atoms are exceedingly small, they contain an enormous amount of energy. This energy can be released and converted into electrical energy for our everyday use.

The energy of the atom

The protons and neutrons in the nucleus of an atom are spinning in much the same way as the earth spins on its axis and yet they are held together in a particular way by a force which is a source of energy. In relation to the size of the protons and neutrons, this force is millions of times stronger than the force of gravity is to us. When the energy holding the protons and neutrons of a single atom is released, it is so weak that it is difficult to measure, but because substances contain so many atoms, their combined energy adds up to an enormous amount. Like adding grains of sand to a see-saw, one grain will have no effect, but if you suddenly poured a big bucketful, your friend at the other end would shoot up into the air with the amount of energy suddenly released.

An atom is the smallest part of an element that can be recognised as that element. But it is possible to break atoms into their smaller particles and it is the number of particles in an atom that makes one different from another.

Opposite: This diagram shows a chain reaction in an atomic bomb, which contains a radioactive metal whose atoms break down to release neutrons. These strike other atoms, which split to form energy and more neutrons, and thus an uncontrollable chain reaction is set up, which gets faster and faster until the released energy causes the metal to explode.

Elements and compounds

Elements are the basic substances from which everything in the universe is made. Ninety-four elements occur in nature, but some are very rare. Some elements can be

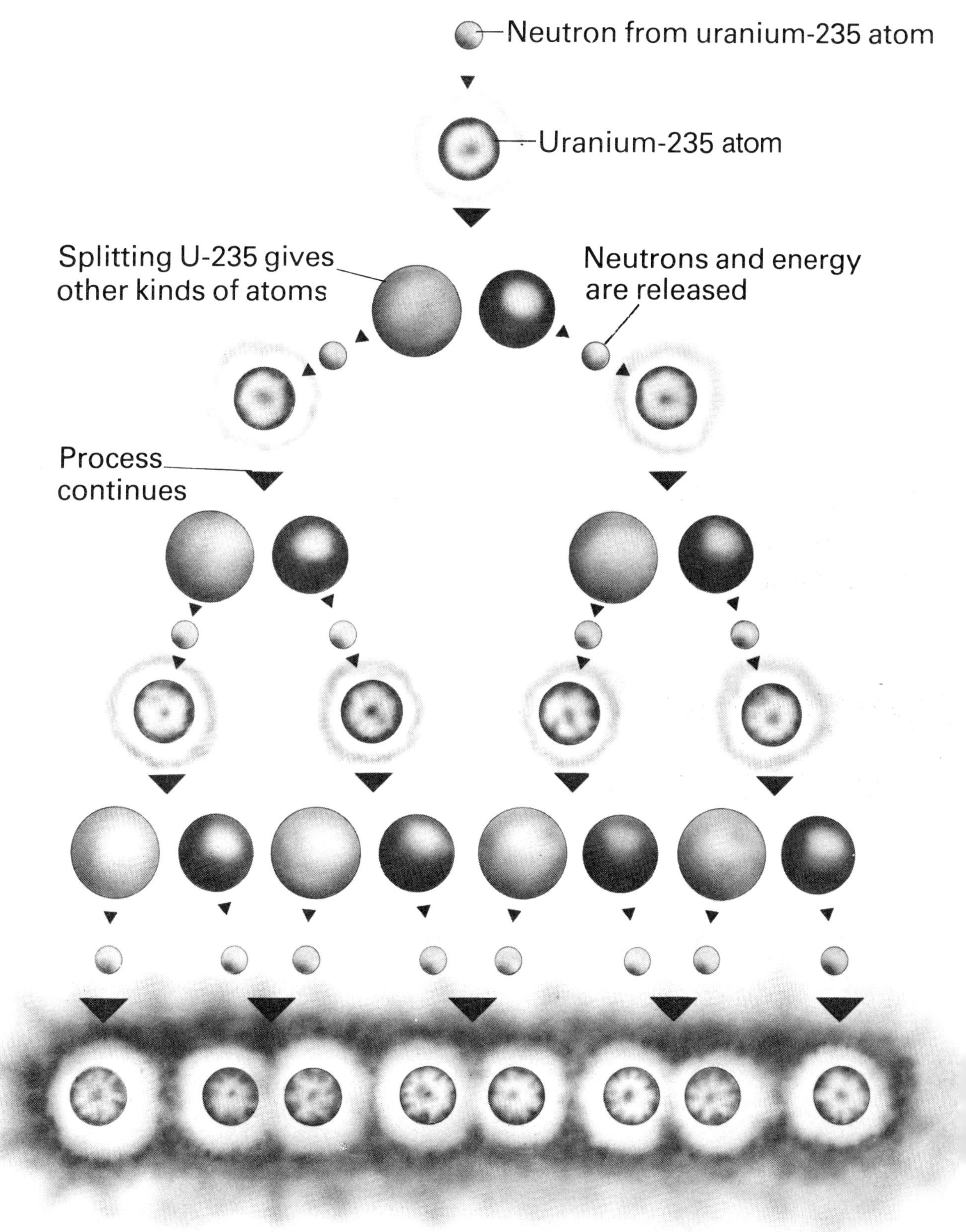

Process gets faster and larger until there is an explosion

Below: Gold occurs usually in the form of a free metal. This soft, malleable and ductile metal is highly prized for its bright yellow colour and it is widely used in jewellery, coins and dentistry. This native gold is in the form of crystals.

Below right: Like gold, silver is a soft, malleable metal, which is often used in making coins and jewellery. This Zambian silver is present on a piece of limestone. The best-known conductor of electricity, silver sometimes occurs as argentite silver, horn silver and other compounds and can be extracted by alloying with lead and separating by cupellation in a special furnace whereby the impurities are oxidised by air.

found singly, but most occur mixed together in *compounds.* Gold is an element that is often found in a pure form, but iron, for example, is usually found as iron ore, which is often iron oxide, a compound of iron and oxygen. Water is a compound of two elements, the element hydrogen and the element oxygen.

Elements such as oxygen and nitrogen are gases at normal temperatures and the element mercury is a liquid, but elements are usually solids. Some elements, eleven in fact, are produced artificially, so the total number of elements is 105.

On earth, oxygen is the most abundant element. It is present in water, mixed with hydrogen; it is present in the atmosphere and it is also present in many minerals.

Some elements are much more suitable than others for converting into energy, and the most suitable ones are usually the radioactive elements like uranium and plutonium. Over time, the radioactive elements change, or break down into other elements, and when some of these elements are artificially broken down, huge amounts of energy can be released.

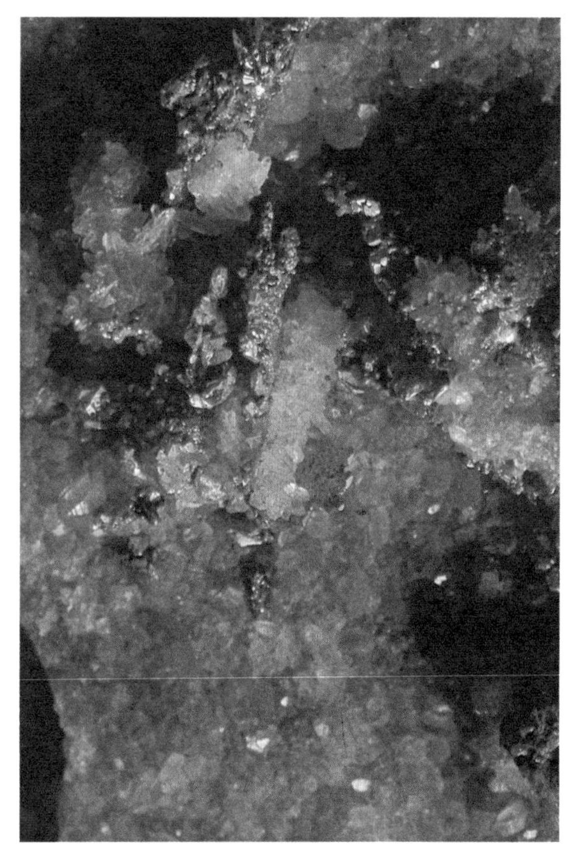

Above Left: This is haematite, the most important iron ore.
A white, magnetic metal, iron can be extracted by the blast furnace method in factories.

Above: Copper is a red, malleable metal and is the best conductor of electricity, after silver.
Copper may also be found as cuprite, copper glance and pyrites.

Uranium

Uranium, in various forms, makes a suitable fuel for converting into energy. Elements can occur in different forms called *isotopes* and natural uranium is a mixture of two main isotopes; 99.27 per cent being uranium-238 and 0.72 per cent being uranium-235. Uranium-235 contains a total of 235 protons and neutrons whereas uranium-238 has an additional three neutrons. Uranium-235 can be used to produce *nuclear energy* and uranium-238 can be converted into plutonium-239 which can also be used to produce nuclear energy.

Uranium was discovered in 1789 by the German chemist Martin Klaproth, but it was not until World War II that the first application of nuclear power was made.

Nuclear fuels are very costly to produce and they are comparatively rare, but they are very efficient. A piece of uranium the size of a table-tennis ball can produce more energy than a thousand tonnes of coal. Also, some

A naturally occurring radioactive element, uranium is a hard, white metal that is of great importance in nuclear reactors. Often found in pitchblende, it is usually coloured yellow or green and is now one of the world's most important and valuable industrial resources.

nuclear reactors actually 'breed' more fuel than they use.

Uranium atoms are comparatively large and in the nuclear process they are made to break apart and form smaller atoms. Hydrogen atoms are much lighter and they can be made to combine. Both these processes produce energy.

Fission

Uranium is a *radioactive* material. This means that some of its neutrons and protons are being lost from it as *radiation.* It is possible to 'bombard' the uranium atoms with more neutrons which can be 'captured' by the charged

This bar chart compares the heat produced by the different fuels. On average, liquid fuels produce more heat than do solid fuels, but uranium-235, which is used in atomic power stations, produces most heat.

Wood

Lignite
(Brown coal)

Bituminous coal

Diesel oil

Gasoline

Uranium-235 (atomic power fuel)

particles within the nucleus of the atom. But the atom of uranium cannot exist with these extra neutrons in its nucleus and it splits apart. When it does, the energy holding the original protons and neutrons together is released and becomes converted to heat. The process is called *fission.* Inside a nuclear reactor is a special liquid, or in some cases a gas, which absorbs the heat and in turn heats a boiler filled with water. The water is turned into steam which then drives an electrical generator.

The chain reaction

Under suitable conditions, the neutrons produced by the splitting of the nucleus can in turn split other nuclei next to them which in turn split more nuclei. This is called a *chain reaction* and it occurs when the right amount, or *critical mass,* of radioactive material is brought together. If the amount is too small, too many of the neutrons produced by fission escape for a chain reaction to continue. An uncontrolled chain reaction causes too many of the nuclei to split at once and a devastating explosion takes place. This is the process of an atom bomb.

Uranium ore is often mined by the open-cut method. In one year, 871,000 tonnes of uranium ore were excavated from this mine. The most important vein deposits are worked in Zaire, Australia, Canada and South America. Uranium also occurs in sedimentary deposits and can be detected using a geiger counter. Once extracted from the ground, the ore is crushed, dissolved in a strong acid or alkali and then filtered. It is recovered by an ion-exchange process, and the resulting yellow cake is made into fuel for nuclear power.

NUCLEAR ENERGY

A *nuclear reactor* is designed to bring together the required critical mass of fuel and provide the right conditions to continue a controlled chain reaction. The part of the reactor where the fuel is located is called the *core.* In one type of core there is a block of graphite with holes drilled in it. The graphite block is called a *moderator* and its purpose is to slow down the faster moving neutrons.

This view of the interior of an atomic power station shows the nuclear reactor (in the foreground) and the fuel-rod charge machine. The technicians are replacing radiation shields on the top of the nuclear reactor.

Reactors

In this atomic power station, atomic energy is converted into electrical energy, which is distributed through a grid to homes and to industry. Massive concrete slabs are used in the construction of atomic power stations as a safety measure.

Uranium fuel is made into *fuel rods* which are inserted in the holes in the graphite block. The fuel consists of uranium-238 and small amounts of uranium-235. Uranium-235 is unstable, it quickly begins to change its structure by breaking down. The U-235 begins to split, throwing off fast-moving neutrons, they have to be slowed down by the moderator. When the slowed down neutrons hit the nonfissionable U-238 atoms, they are absorbed and plutonium-239 is made.

To stop the chain reaction getting out of control, a number of other rods, called *control rods,* are also placed

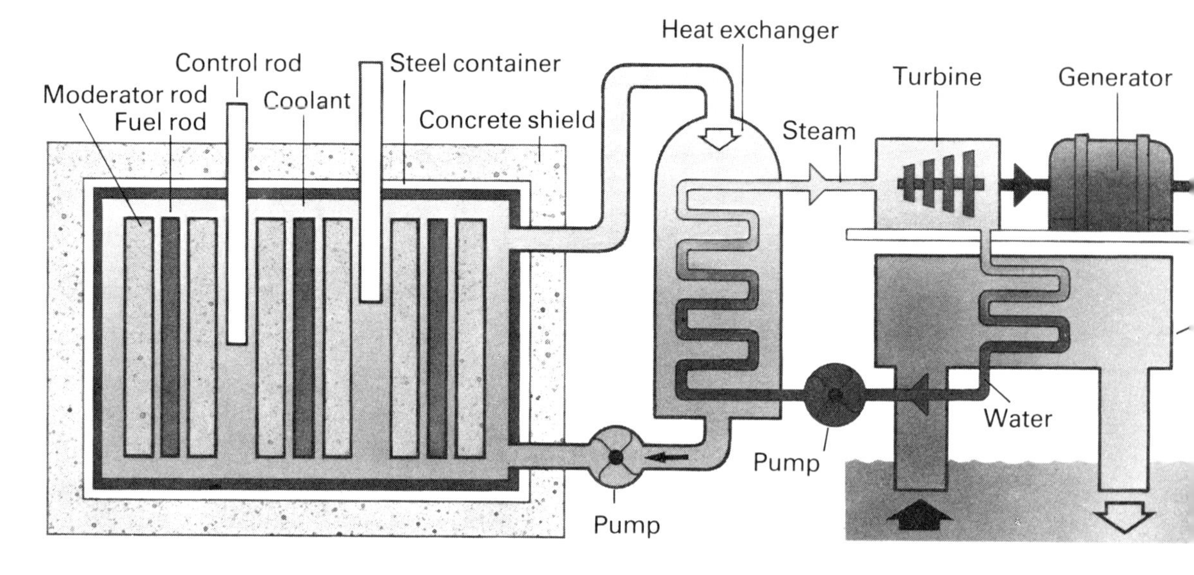

The fuel rods in the reactor contain uranium-235 atoms, which split apart and set up a chain reaction releasing large amounts of heat. The heat is ducted away from the rods into a heat exchanger, where it is used to turn water into steam. The steam drives a turbine, which is coupled to a generator, and the electricity produced is distributed to factories and homes through the grid system.

in the moderator. These are made from substances which have the ability to absorb neutrons, so if there are too many neutrons being released at any one time, the rods will soak them up. When these control rods are pushed all the way in, the reactor can be shut down completely. The whole system of moderator, fuel rods and control rods is sometimes referred to as an atomic pile.

Types of reactors

The commonest reactors use the process of slow neutron fission controlled by a moderator. They are called *thermal reactors.* They are usually classified by the kind of coolant, or moderator, used. This can be boiling water, pressurised water, heavy water or water containing an isotope of hydrogen, gas, or graphite block. The fuel may be natural uranium or it may be uranium oxide, treated or 'enriched' so as to increase the amount of U-235.

A different, and important, type of reactor is the *fast-breeder reactor,* which breeds more fuel than it uses. Its fuel is uranium oxide highly enriched with U-235 or plutonium-239. Surrounding the enriched fuel is a 'blanket' of U-238. Fission takes place in the enriched fuel, gradually using it up. At the same time, more plutonium-239 is produced as neutrons are absorbed in the surrounding U-238. The newly-bred fuel can be extracted and used to enrich the core again. The fast-breeder reactor needs no moderator and can therefore be smaller than a thermal reactor.

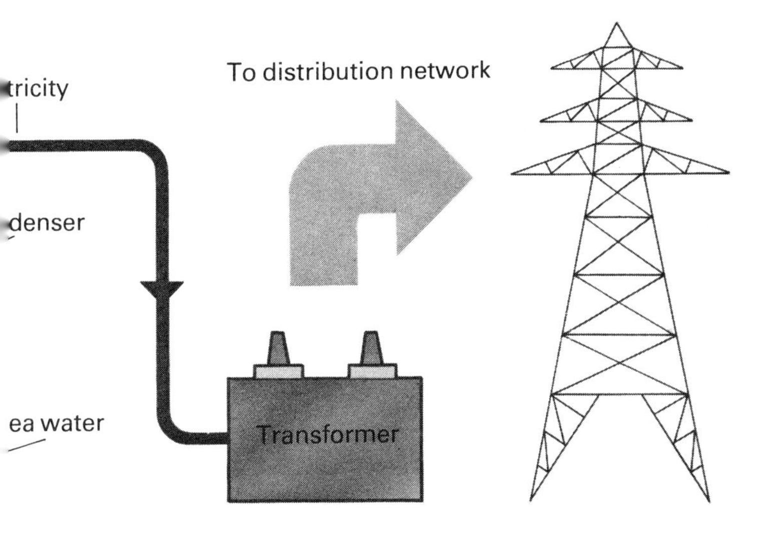

Fusions

Under the right conditions, the nuclei of two atoms can be brought together to produce a larger atom. When this happens, the new atom is not quite equal in mass to the combined masses of the original two atoms. This means

These men are working on the fuel element canning line at one of the British Nuclear Fuels works that manufactures uranium metal fuel for use in big nuclear power stations in Great Britain and overseas.

This is a fuel rod assembly for an advanced gas-cooled nuclear reactor. Each fuel rod, containing an enriched radioactive fuel, is capped after the fuel is inserted. The complete rod assembly can be lowered automatically into a nuclear reactor by means of a fuel-rod machine, which inserts new rods and removes spent ones.

that the amount of energy required to hold the different masses together is not the same as before, but is, in fact, less. Some energy, therefore, is released.

The process occurs when two atoms of deuterium (an isotope of hydrogen) are combined to produce helium. To achieve this, extremely high temperatures of several million degrees C are required. The hydrogen bomb uses deuterium and tritium and to get the high temperatures required it is triggered off by an atomic explosion. To produce energy by the fusion process for commercial use has so far proved very difficult.

Dangers of nuclear power

An atomic bomb is a nuclear reaction specially designed to release all of its potential energy in a fraction of a

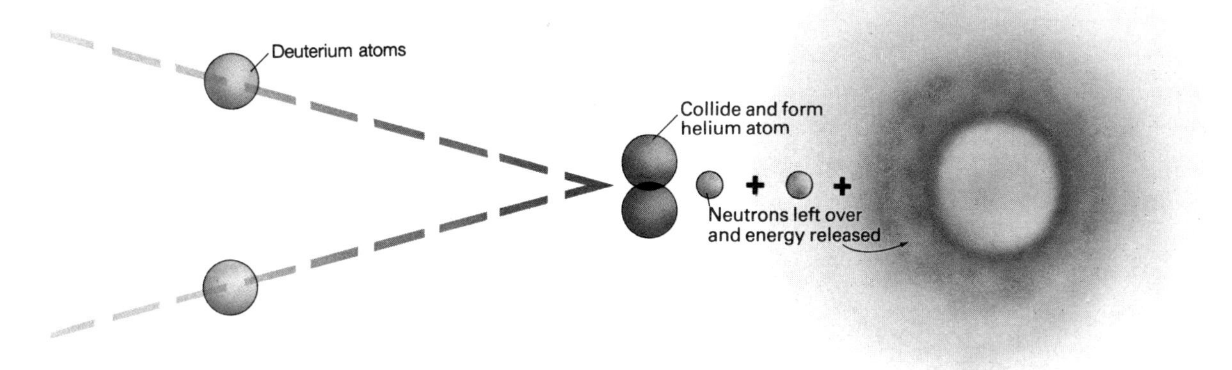

When two deuterium atoms collide, they combine to form a helium atom, neutrons and a large amount of energy. Thus nuclear fusion occurs when two light nuclei collide to form a heavier nucleus, releasing both heat and energy.

second. It will destroy all living things and knock down buildings many kilometres away.

In an atomic power station, scientists are working with atoms they have deliberately made unstable and, without extreme care, the atomic pile could possibly exceed its critical mass and become a potential atomic bomb, with disastrous consequences.

Radiation

When the nuclei of atoms break up, they produce *radiation.* Radiation is energy given out by the atoms and may consist of particles called alpha particles and beta particles, or it may be in the form of electromagnetic waves, known as *gamma* rays. Sometimes, and in controlled amounts, radiation can be of great benefit, but too much is harmful and produces what is known as radiation sickness. Scientists working the atomic pile have to be protected from the direct radiation that escapes during the nuclear process by a thick wall of concrete and steel. They cannot be in direct contact with the rods of radioactive material and use remote control mechanical 'arms and fingers'. When it is necessary for them to be able to see the materials they are working with, they are protected by a special window of lead glass.

Atomic waste

Both the fuel rods and control rods contain unstable atoms, even after they have served their time in the reactor. So does all of the other material used in the process. This includes the protective clothing worn by the scientists and technicians, and worn-out machinery which has been in close contact with the atomic pile. This material all becomes radioactive and can cause sickness in anyone who is exposed to it for any length of time. It will also make other substances and living things radioactive. It is called atomic waste and special care must be taken with its disposal. It is often placed inside steel drums and buried underground or dumped at sea.

The danger from atomic wastes can last hundreds of years until the contaminated substances lose their radioactivity, and this is one argument many people have against wanting to establish any more atomic power stations. Should the steel drums in which the waste is stored begin to leak, a number of things may happen. If the drums are in the oceans, then the fish could become poisoned. If the drums are on land, the grass could become poisoned, the cattle would eat the grass and when we eat meat, or the fish from the sea, the radioactive poison would be passed on to us.

Atomic war

Another problem with the manufacture of nuclear power is the use of similar processes and materials in atomic warfare. Countries which are able to use nuclear materials

This sinister-looking, mushroom-shaped cloud occurred when a thermonuclear device was exploded in the Marshall Islands by the United States in 1952. As a result, the test island of Elugelab was completely destroyed. After the signing of the Test Ban Treaty of 1963, it was prohibited to test any nuclear weapons in the atmosphere or underwater.

for peaceful purposes can also use similar materials for making atomic bombs. As more and more countries are able to manufacture nuclear power, so, potentially, more could use this knowledge for aggressive purposes. This is one of the main objections of many people who are opposed to the expanding use of nuclear power in the world.

SOLAR ENERGY

The sun is believed to produce its energy by the fusion process, turning hydrogen into helium in a tremendous continuous reaction and pouring out huge amounts of energy day after day, year after year. Sunlight is the visible evidence we have of the sun's output of electromagnetic radiation. When we suffer sunburn, we experience another effect of the sun's powerful radiation. In spite of the enormous distance between the sun and earth, the radiation put out by the sun is very powerful. We are shielded from harmful effects by the atmosphere which surrounds the earth and particularly by the ozone layer high above the earth.

This cross-section of the sun shows how heat travels from the central core to the surface. Nuclear transformations in the solar interior create energy, which keeps the sun radiating heat and light. Hydrogen is the principal constituent in the production of solar energy and, when eventually the hydrogen starts to become exhausted, the sun will change its structure.

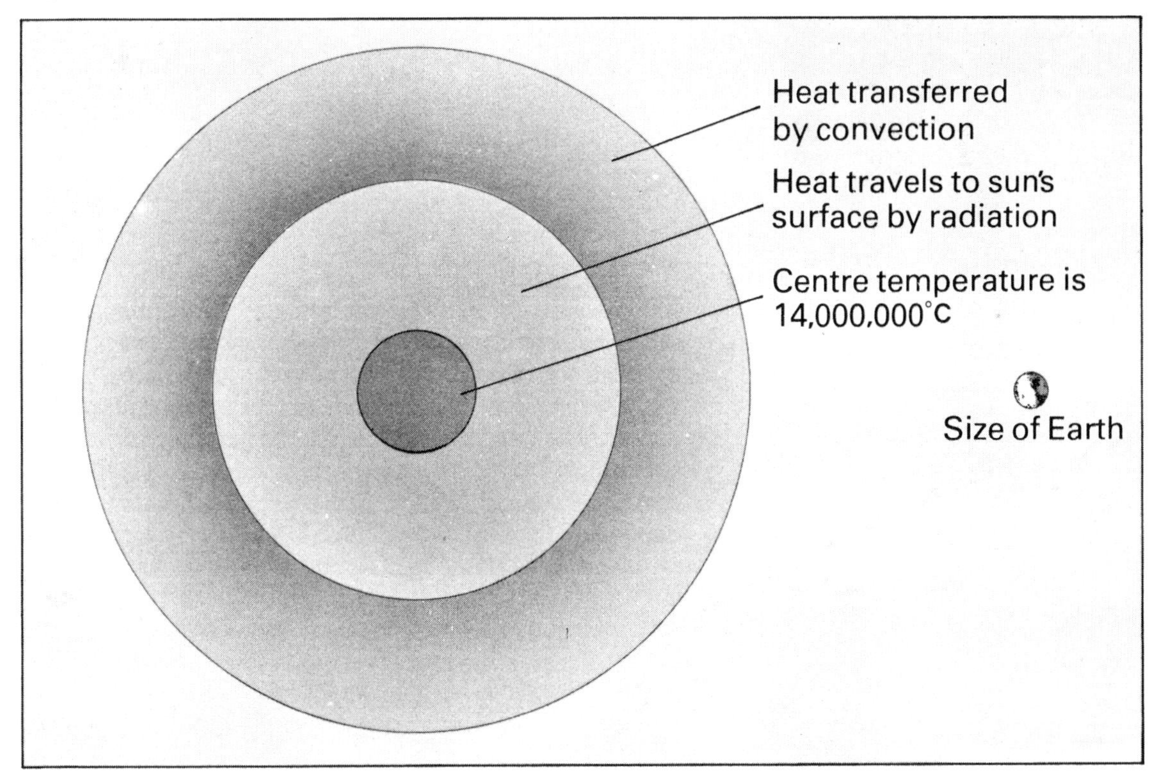

Even a huge greenhouse can be adequately heated by the sun. Solar radiation is transmitted by the glass windows and most of the reradiated infra-red radiation from within is returned to the greenhouse. The temperature inside is thus higher than that outside.

The sun is the source of all energy and this diagram shows the greenhouse effect. The glass panes of the greenhouse are penetrated by short wave solar radiation to warm the surface below and create warmth inside the greenhouse. The glass reflects the long wave radiation from within the greenhouse.

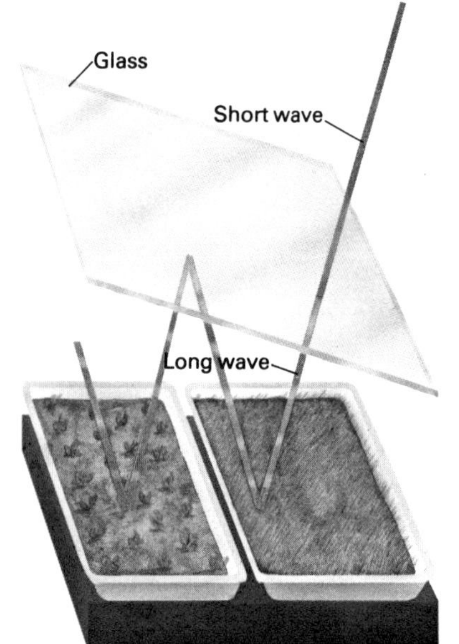

We know that the sun is responsible for all life on earth and that it provides the plants with the heat and light they need for growth. Now, with the fossil fuels in short supply, scientists are looking at the sun to provide supplies of energy for our everyday needs in other ways. One of the easiest ways we can use the energy of the sun to provide energy is in the form of solar heating to keep our homes, factories and offices warm in winter.

The greenhouse effect

The basis of many of the solar heating devices being used today is what is known as the 'greenhouse' effect. This is what happens when sunlight passes through glass into a small, enclosed area. The most common example of the greenhouse effect is the family motor car. You will have experienced the effect on a hot day when the windows have been left wound up while the car is in the sun. When you open the doors, a gust of very hot air rushes out. This air, having been trapped inside the car, has got hotter and hotter.

The greenhouse effect is possible because of the amazing properties of glass in relation to sunlight. When sunlight passes through a glass window, the heat in the sunlight also passes through but the glass traps the heat behind it. While the outside temperature remains high during the day, the heat from sunlight is built up behind the glass. As more heat passes through the glass, the air inside is warmed up more. Even in the winter, there is enough heat in the sun's rays to make the air behind the glass of a greenhouse quite hot.

The energy conversions in both steam and water turbines can be traced back to the same source: the sun. The sun's radiant energy produced the oil and coal needed to power a steam turbine. The mechanical energy produced is then converted into electrical energy. The sun's potential energy causes water to evaporate from the sea and lakes to form rain clouds. When it rains, the rivers and lakes are replenished and their kinetic energy is used to drive a large water turbine.

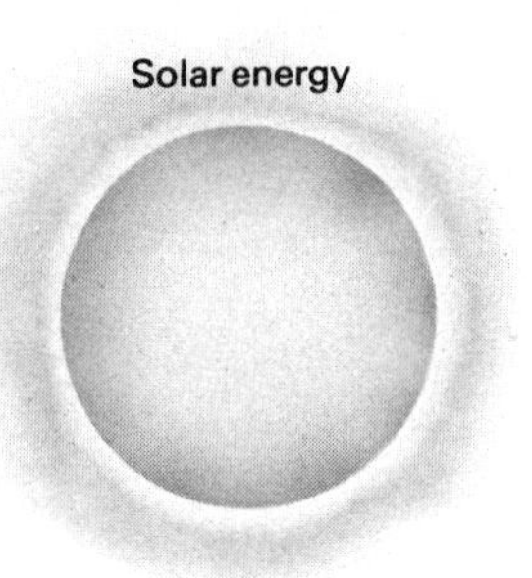

Solar energy

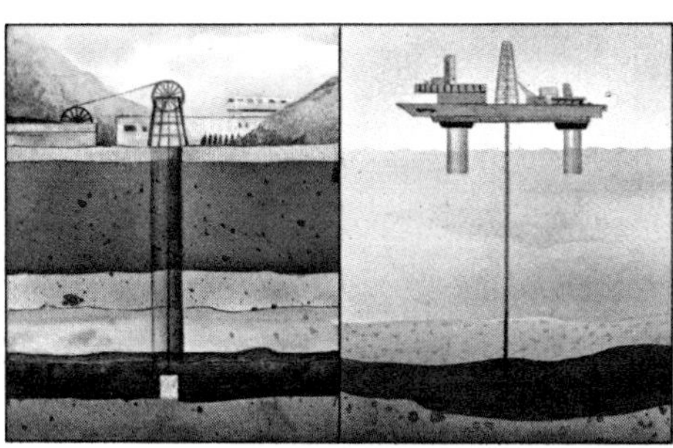

Coal Oil

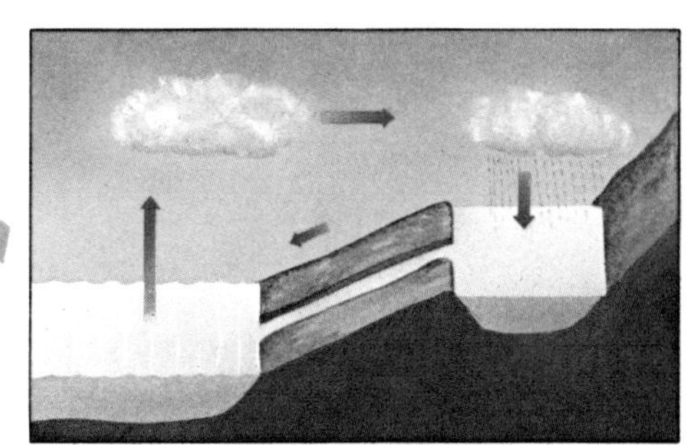

Hydro system

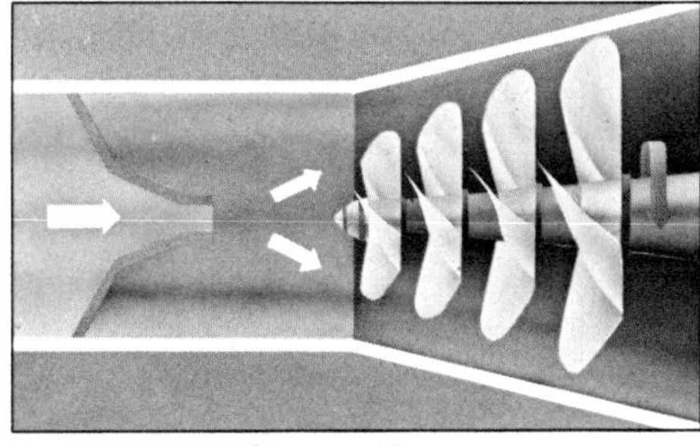

Steam turbine

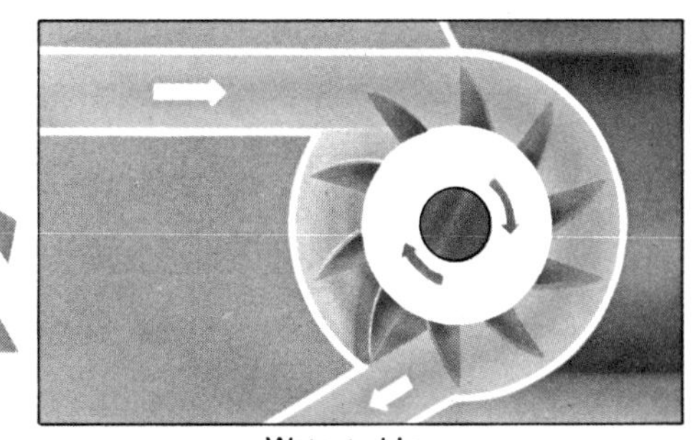

Water turbine

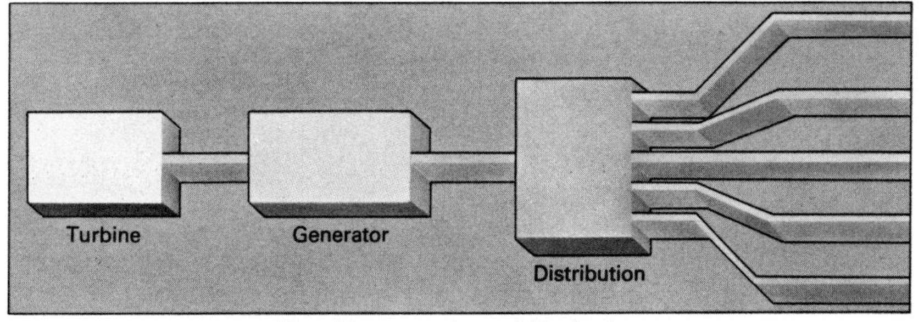

Electricity generation and distribution

Solar energy collectors

Opposite: Flat-plate solar collectors absorb any unfocused sunlight, which, in turn, heats the water circulating inside the collector. The warmed water is carried to a heat exchanger in an insulated hot water tank. Solar energy can thus be harnessed to heat people's homes both cheaply and efficiently.

A *solar collector* is the name given to a device that uses the greenhouse effect to concentrate the heat of the sun's rays. The rays are falling with the same intensity, for example, on every part of your house during the daytime. To heat up a whole room using the greenhouse effect is very inefficient, unless the glass area is very great in proportion to the size of the room. But if you make a small box, with a glass top to let the sun's rays through and insulated to keep the heat inside, then you are able to heat the small amount of air inside the box to very high temperatures. Most solar heat collectors work on this principle.

A solar collector may be a shallow box made of wood, metal or plastic. The bottom and sides of the box are painted black, or lined with black plastic. Black absorbs heat, holding it inside the box instead of reflecting it away. To prevent loss of heat, the bottom of the box is lined with insulating material, such as glass fibre. The box is fitted with a glass top, which can be a single sheet of glass, or it may be double glazed, that is, two sheets of glass with a thin space of air between them.

Governments are investigating the harnessing of solar energy as a prospective means of heating people's homes and water supply, both cheaply and efficiently. These solar panels are being tested at the National Centre for Alternative Technology at Machynlleth in Wales.

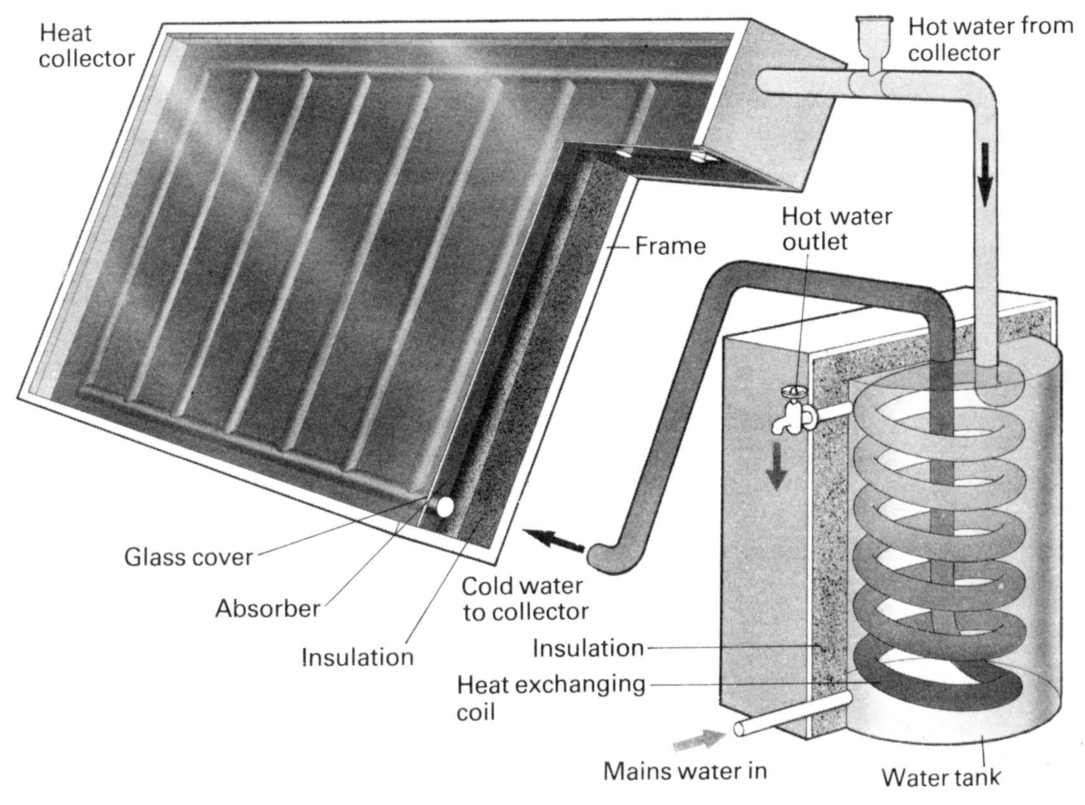

Transferring the heat

The air that is warmed up inside the solar collector can be used to heat water. This is done several ways. One of the simplest methods is known as the *trickle system.* In this system a sheet of corrugated metal is placed inside the box and water is fed through a pipe at the top. The pipe has small holes at the bottoms of the corrugations. The metal is warmed up by the heated air and sunlight, the water trickles slowly down the corrugations in the metal and gets warm. The warmed water goes into a pipe and into a storage tank. By circulating over the corrugated metal throughout the day, the water becomes hot enough to provide most of a family's usual hot water needs.

A more efficient system uses metal tubes, such as copper, to circulate the water inside the solar collector box. Instead of corrugated metal, a flat sheet of metal lies at the bottom of the box and a copper pipe is bent in a

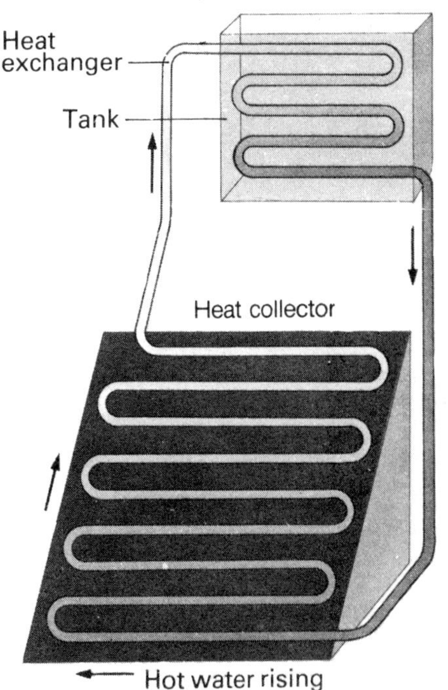

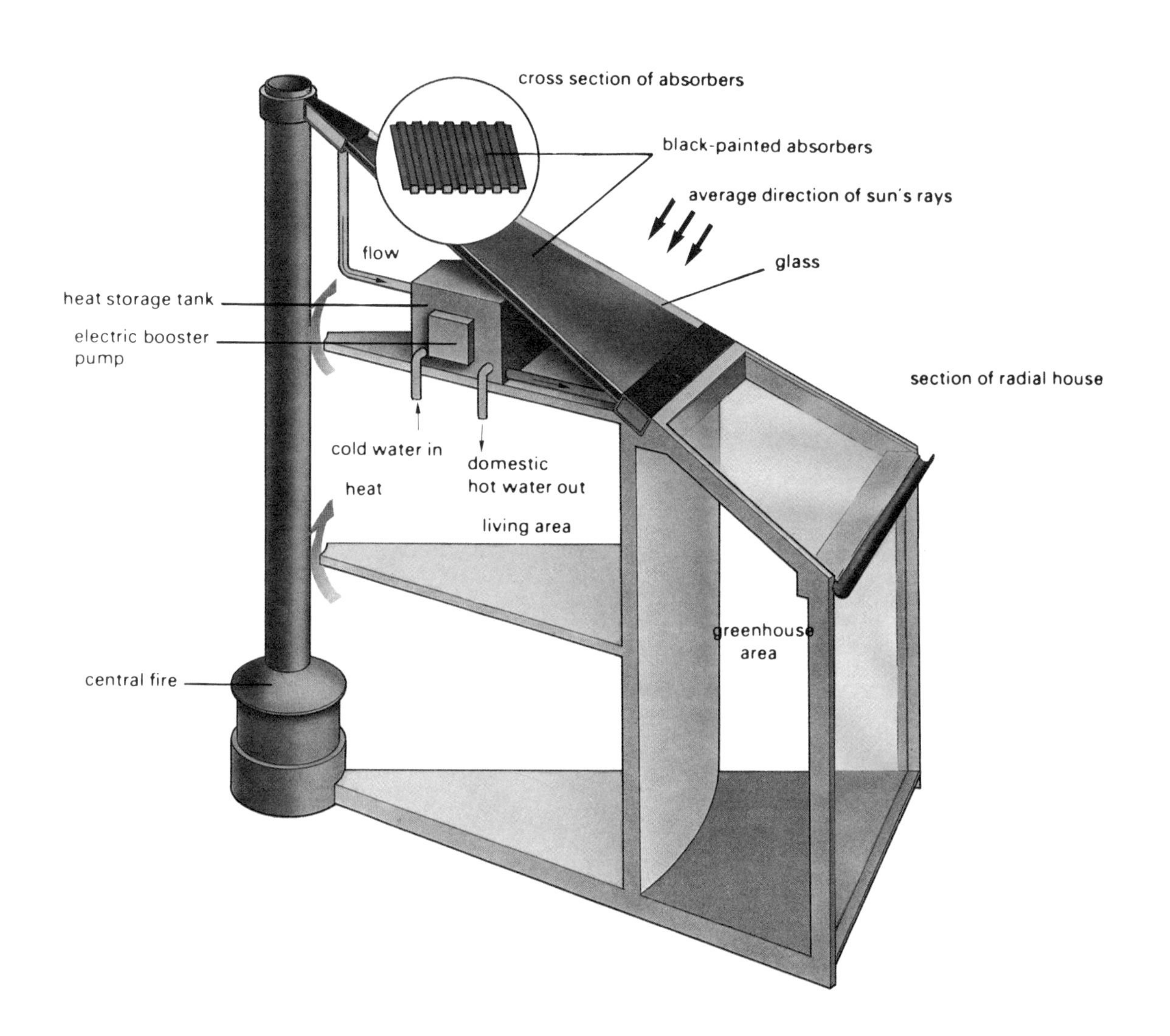
cross section of absorbers
black-painted absorbers
average direction of sun's rays
flow
glass
heat storage tank
electric booster pump
section of radial house
cold water in
domestic hot water out
heat
living area
greenhouse area
central fire

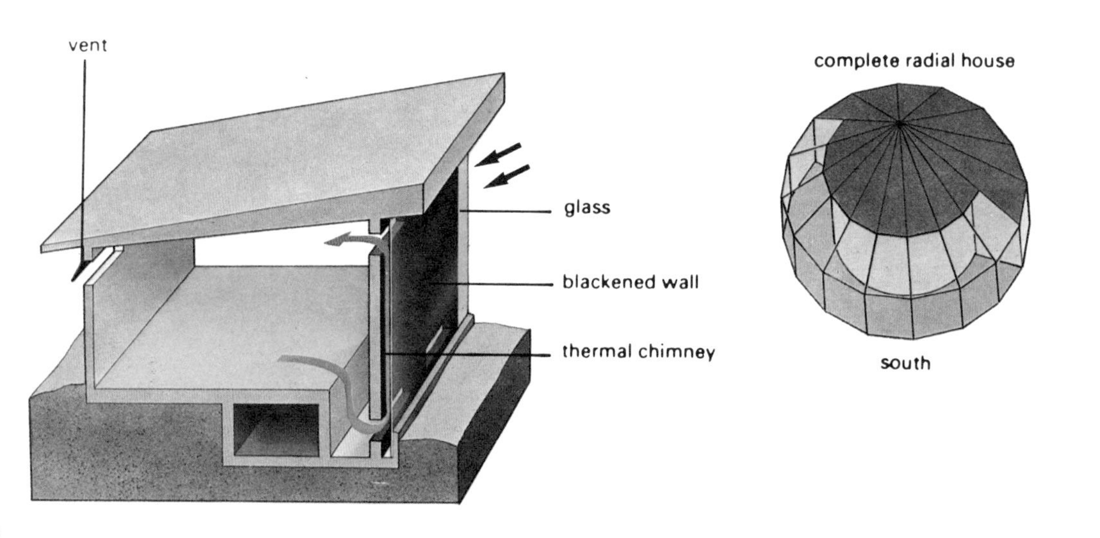
vent
complete radial house
glass
blackened wall
thermal chimney
south

series of U-curves and soldered to the metal sheet. Water comes in at the top, flows around inside the copper pipe and passes out at the bottom. The heat absorbed by the metal plate passes through the copper and heats the water.

In both these systems and in fact in most solar heating systems, the water is stored in an insulated tank similar to the tank used for gas or electric heating. It is estimated that in Australia about 60 to 95 per cent of a family's hot water needs could be obtained from this type of solar collector, winter and summer.

Specially designed plastic, metal and glass collectors are now being designed and manufactured.

Heating the house

Water heated by the sun during the day can also be used to heat the house at night. This is best done by having a concrete floor in which water pipes are embedded. At night, the hot water from the storage tank is circulated through the pipes in the floor. Heat from the hot water is transferred to the concrete and radiates up into the room.

Another way of using the sun's energy for heating is to heat air in a solar collector and let it circulate inside the house. This is not as efficient as using heated water

Opposite page:
The experimental radial house uses solar energy for hot water and central heating purposes. The section through the radial house shows the inner living area and the outer greenhouse for growing plants. Hot air enters and leaves the rooms by a system of vents in the walls.

Below: One of the most advanced and sophisticated experimental solar houses is situated in New Mexico in the United States. During the daylight hours, a wall opens to allow the sun to heat the special solar panels, which, in turn, heat the water in the house. At night the wall closes, and it is thought that this solar heating design is far more attractive than are solar collector panels set into the roofs of houses.

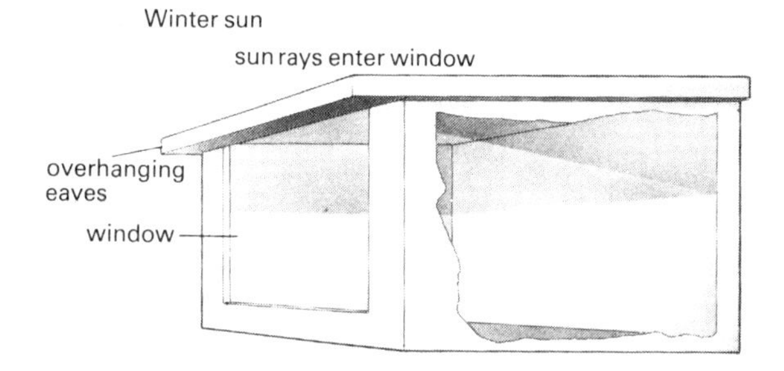

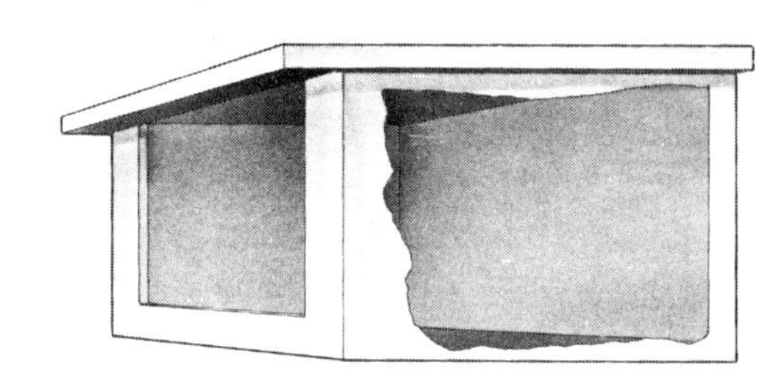

In winter, when the sun is much lower in the sky, the rays warm the solar house by shining straight through the windows

However, in summer, when the sun is higher in the sky, the house's large, overhanging eaves prevent too much sunshine entering the windows, and thus the house does not become overheated.

because the air does not hold its heat as long as water but when a house is specially designed as a 'solar house' air heating becomes more efficient.

The solar house

Great use can be made of the sun's warmth to heat up water, as in a solar collector, air for circulating inside the house, and the house itself. Concrete is a substance which absorbs heat, holds it and then releases the heat later on. A solar house, therefore, will most likely have a concrete floor. Over the concrete may be placed tiles which also absorb and then radiate back heat.

Another important factor in the solar house is the position of the sun during the winter and summer months. In winter, the sun is low in the sky and may shine more or less straight through a window. A solar house has as many windows as possible facing the winter sun to help heat the house. The sun warms the air in the rooms and this air can be circulated to the rest of the house. It also falls on the tiles and heat is absorbed into the concrete floor. At night, the warmed floor will radiate its warmth back into the rooms. When the floor has pipes for hot water embedded in it, it will be warmer still. A feature of this type of solar house is the absence of carpets, because they would insulate the floor from the sun's rays and stop the heat getting through. Scatter rugs are used instead.

In summer, the sun is higher in the sky and a solar house will have large eaves to stop as much of the summer sun as possible from shining in through the windows. In this way, the house is not heated directly by the sun's rays in summer. Insulation in the roof, which should be provided to keep warmth in in winter, will also help keep out heat from the roof. The house is designed so that any natural breezes may pass right through, removing warm air that collects in the rooms.

Once again, the concrete floor is important. At night, it cools down and will keep cool for most of the next day. It is also possible to run cold water through the pipes in the concrete to help keep the house cool. By shading the windows during the summer it is possible to keep the inside of the house much cooler than if the windows are left exposed.

The methods of solar heating and cooling are very simple, but they can reduce the amount of fuel needed by the average house by as much as 90 per cent. In the past, houses have been designed without much regard to using solar energy, largely because we relied on electrical energy to provide heating when we needed it and cooling in the summer. But with energy supplies dwindling, more and more houses will be designed and built as 'solar houses'.

This experimental solar house in Arizona, United States, is set in the heart of the dry desert. The solar panels are tilted at an angle to the hot sunshine. In desert areas and open country, the atmosphere is clearer and therefore the intensity of solar radiation is much greater.

POWER FOR THE FUTURE

As well as providing comfortable living conditions such as those provided by the solar house we must have other ways to provide energy to run factories, transport, television stations, cinemas and all the appliances in the home. Most alternative power sources depend, in the long run, on the energy of the sun.

The solar collectors described for use in the solar house can provide home heating and hot water, but for greater energy output more efficient solar collectors are needed. In one new type, developed by the University of Sydney, very high temperatures can be gained.

In this type of solar collector, glass tubes are used instead of metal tubes. The tubing is quite complicated, each unit consisting of one set of glass tubes inside another set. There is a vacuum between the two sets of tubes which increases the heating capacity of the collector considerably, by greatly reducing loss of heat. Another refinement is what is called a *selective surface.* This is a special coating applied to the tubes which absorbs a high percentage of the sun's heat without re-radiating much of it away as a heat loss. In one

This simple bar chart shows the energy consumption of some of the world's most advanced industrial nations. The natural oil supplies of the Persian Gulf, United States and Soviet Union are rapidly diminishing as the demand for oil increases. Therefore it is more important than ever before that the governments of industrialised consumer societies investigate alternative sources of energy. Solar, wave and tidal energy, wind power and geothermal energy and synthetic fuels form current areas of research.

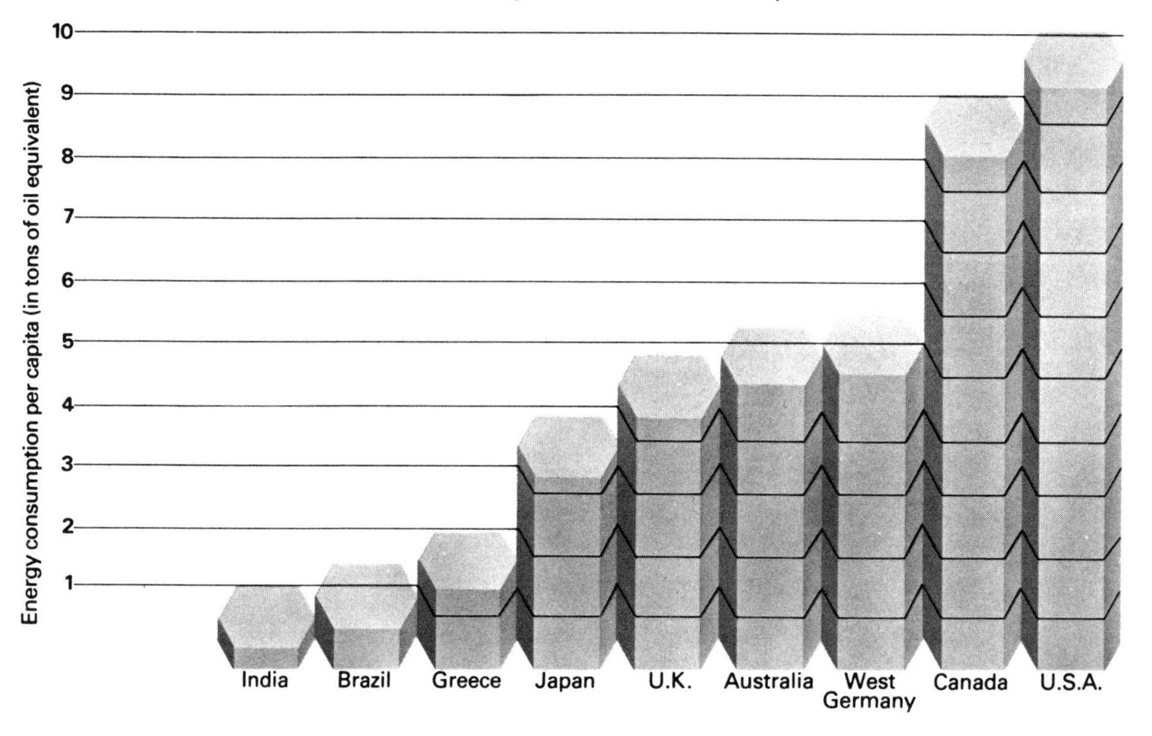

This spectacular solar furnace is situated at Odeillo in the French Pyrenees. A north-facing parabolic reflecting surface is formed by 9,000 mirrors, and a further 11,000 flat mirrors on the opposite hill can be adjusted to direct the sunlight on to the glowing reflector. The furnace room is located in the centre of the reflector, and the temperature inside the furnace can rise to 3,800° C.

experiment the University used oil instead of water inside the glass tubes and heated the oil to 300°C, hot enough to generate steam and, in turn, drive a generator.

With collectors of this efficiency, once they are fully developed, it could be possible for every home to have its own solar power supply mounted on the roof. The same could apply to factories and office blocks.

Solar furnaces

Solar furnaces are being developed for the large-scale production of energy from the sun, such as the generation of electricity for commercial and industrial use.

A solar furnace is like a huge concave mirror which concentrates the sun's rays onto one small area. The concentration of sunlight increases the heat output so much that, in one solar furnace in France, the temperature reaches 3,800° C.

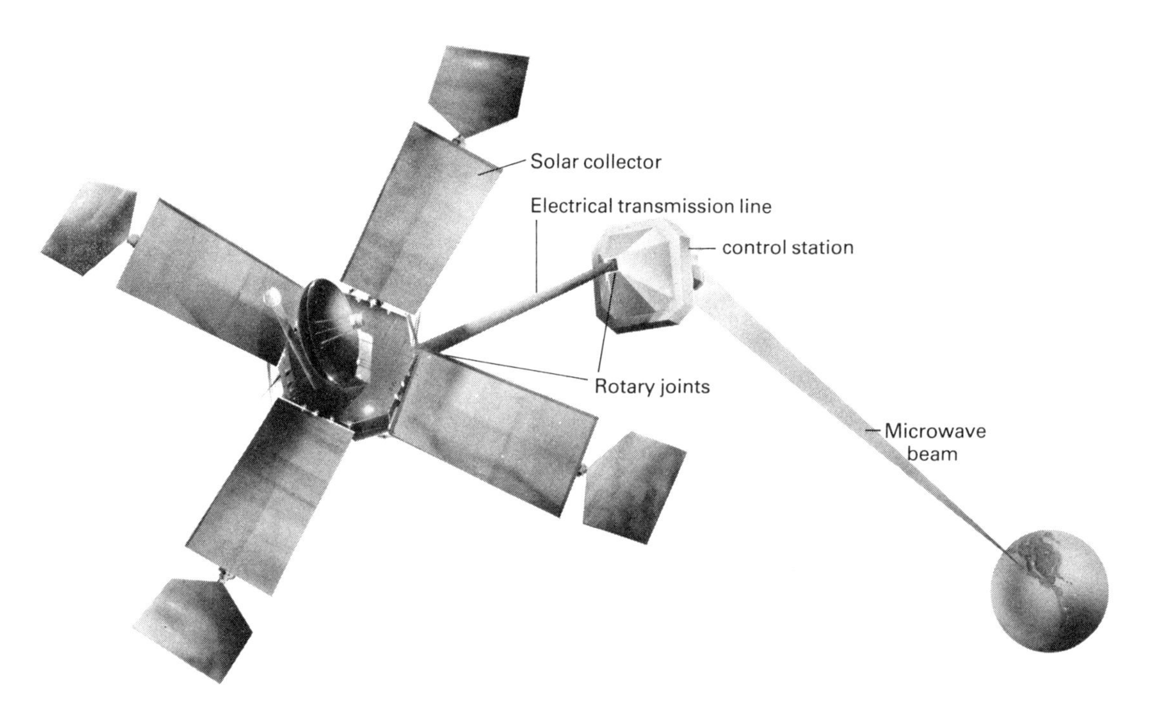

This solar collector is still speculative only, although solar collector panels are being used in communications satellites. The principle is that solar energy would be collected on special collector plates and converted within the satellite itself to microwave energy, which is beamed down to a receiving station on earth. Scientists could maintain contact with the satellite and send it precise instruction signals.

Solar cells

A solar cell is a device that can turn sunlight into electricity. Solar cells are used in some photographic light meters and they are also used in spacecraft. They can be made of thin wafers of silicon treated with special chemicals. When sunlight falls on them, electrons are made to flow through an electric circuit. As well as in spacecraft and light meters, solar cells are also being used to provide booster current for telephone lines. This is being done on long distance telephone lines in central Australia.

Solar cells have one major drawback and that is cost. With present technology it would cost several hundred dollars just to light one household bulb by using solar cells. However, many people believe that in time more efficient and less expensive solar cells can be produced. This may mean cheap energy for everyone.

Space power stations

Probably the most effective way of producing electricity directly from solar cells will be to locate the cells in space, where they will receive almost continuous sunshine and will not be interfered with by the weather. In a small way,

Silicon cells are used to control the lens aperture of a camera. The element silicon can convert some solar energy into electricity, but the process is still prohibitively expensive. Although silicon accounts for about 25 per cent of the earth's crust, it does not occur naturally in a pure form, and commercial silicon is obtained by reducing silica with carbon in an electric furnace.

this has already been done, for it is the method that has been used to power the instruments of space probes and satellites.

At the present time the process is extremely expensive. The material most used for these cells is silicon, which will convert about 15 per cent of solar energy to electric current, but the cost of electricity production in this way is still about forty times the cost of producing it from coal or uranium. However, other materials are being used for experimental cells, and if these are successful, it is possible that solar powered stations will be mounted in satellites far above earth's surface.

The energy produced would be beamed down to earth in the form of microwaves.

Wind power

The windmill is the most common example of power from the wind. It is used in the country mainly to pump water for irrigation and livestock. In the past, however, windmills were used to grind corn and to drive machinery for sawing timber and making paper. A windmill may have its blades designed in such a way that they will operate in all kinds of winds and actually turn away from the wind when the wind is strong enough to cause damage. Such windmills are not very efficient for the generation of elec-

This windpump and storage tank on an Australian farm need dependable winds for regular and efficient operation. Mounted on the top of a tall steel-lattice tower, the windwheel has pressed-steel blades which are secured to a windshaft. The head rotates around the tower and requires little maintenance. The sprung, hinged tail ensures that the wheel is always turned into the wind and prevents the wheel from racing in strong winds.

tricity, so for this purpose a different design is needed and such machines are generally referred to as *wind generators.*

Wind generators

Small wind generators are now being used by boat owners and campers to keep their batteries charged. This in turn gives them energy to operate the starter motor or to use electric lights at night. A number of different types of wind generators are available for household use, mainly for farmers, to provide electricity to run the lights, refrigerator, vacuum cleaner and TV set.

While farmers and people living in small towns could possibly have their own wind generators, it does not seem likely that city dwellers would be able to use them. Wind generators must have an uninterrupted flow of wind to operate successfully and in cities, with their large number of tall buildings, this would be virtually impossible.

It may be possible to have giant wind generators located outside the cities, generating power which could then be fed into the electricity supply lines. There are certain parts of the world, such as off the coast of South Australia, where powerful winds blow all the time. Wind generators in these 'wind corridors' may be capable of producing electricity day and night. There are many technical problems to be overcome before such a scheme could be made to work, for the generators need to be an immense size to supply enough electricity for a city. Another problem is that, in the wind corridor off the South Australian coast, for example, the wind generators would have to be in the ocean.

Unlike the well known windmill, the blades of future wind generators are likely to be shaped more like those of the propeller of an aeroplane. This design gives better efficiency at high speed. Other designs are being experimented with, such as a curved blade which catches wind no matter which way it is blowing.

It is estimated that by the year 2000, the USA may be using wind power for as much as 5 to 10 per cent of its total energy needs. Generation of energy from the wind is, however, quite complicated. Because of the high speeds that winds can reach during a storm, wind generators must be extremely solidly constructed. The gears needed to drive the electric generators will have to stand up to great strains. These are problems the scientists are still working on, but the day may not be far away when large

Below left: Traditional windmills have numerous small sails. The frames are covered with canvas sails like those of a ship.

Below: This sophisticated aerogenerator was designed by Sir Henry Lawson-Tancred and stands on his estate in England. Wind power is harnessed by the aerogenerator to produce electricity on a small scale. The main objections to its use in rural areas have been environmental, as it is felt that the design is too ugly and conspicuous in areas of great natural beauty and charm.

scale generation of electricity from the wind becomes a reality. Wind power offers a number of advantages. It is non-polluting and it is free, so long as the wind blows.

The Cockerell raft is one of the latest design concepts in the development of machines to harness wave power. Shown on test in the Solent, England, the raft draws in sea water as it rises and falls and, at the same time, water is expelled from another chamber through a low-pressure turbine, which is coupled to a generator.

Wave power

Another non-polluting source of energy is wave power. Every day, the oceans of the world build up a lot of energy which at present is not being used. If you stand on the beach during a storm, you will see how the waves pound against the rocks with enormous force. A number of ideas have been put forward as possible ways to make use of all this power. One of these is to use a series of devices like rafts placed out to sea so that they will rise and fall with the swell of the waves. This form of energy can be converted into electrical energy by means of rams or pistons.

Another scheme is to use the rise and fall of the tides in estuaries. In an estuary, the tide can rise and fall by many metres and it may be possible to use this to drive turbines in a similar way to hydroelectric power schemes.

Left: The oscillating water column is another device for making use of wave power. The water column is secured by piles to the sea bed and, as the waves approach, the column oscillates and causes surface air to be forced through the air turbine.

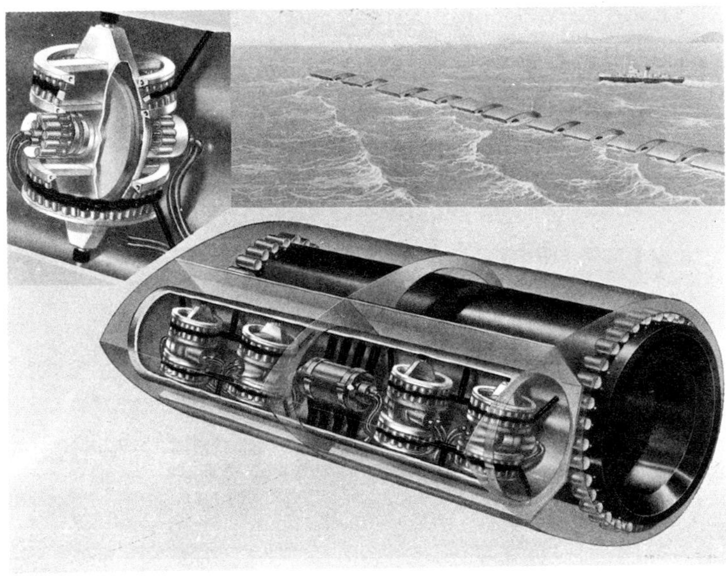

Below left: The mobile Salter duck is oscillated by the wave motion about its spine. The rotary motion of the duck's four gyroscopes causes the multiple cylinders to interact with the cam ring. High-pressure hydraulic fluid is generated to drive the motor and electric generator.

Below: This aerial photograph pinpoints the part of the River Severn, England, that is being considered as a possible location for a tidal barrage in order to harness the tidal power of the estuary and of the Bristol Channel.

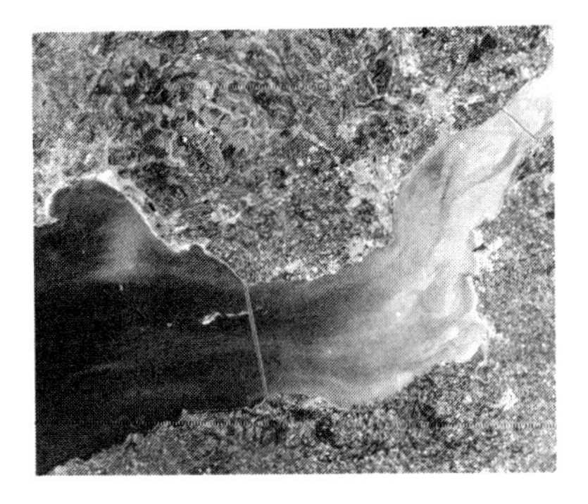

Tidal power

Several tidal power plants already exist. One on the Rance River in France is the only known tidal plant that is producing power for commercial use at present and it has been operating since 1966.

Russia has at least one experimental plant and is believed to be studying the possibility of setting up several large power stations of this type.

Wood

Wood has been used as fuel for centuries, but the problem with wood is that it takes a tree many years to grow and just a few hours to burn. One idea is to locate a factory in the middle of a forest. Trees would be cut from one part of the forest to supply power for the factory. As soon as the trees are cut new ones are planted and so by the time the trees are cut all round the factory, the new ones would be big enough to use as fuel. The success of this idea depends on rapid growth and scientists are trying to breed trees that will grow quickly and still supply good wood for burning.

Methane gas

Methane gas is one of the biofuels, that is, it comes from natural sources such as vegetable and animal waste. It is produced naturally in swamps but it can easily be made in what is called a digester. Methane burns easily. All kinds of waste material such as grass cuttings, cabbage leaves, banana skins, manure and sewage, are fed into the digester. Bacteria starts to break the waste matter down

You can collect methane gas by performing this simple experiment. All you need is a wide-necked jar, filled with water, and a stout stick. Find a swampy or exceptionally muddy area that is filled with rotting vegetation. Prod the mud with a stick and invert the water-filled jar above the spot. When the mud is disturbed, the methane will rise to the surface in bubbles and is caught in the jar.

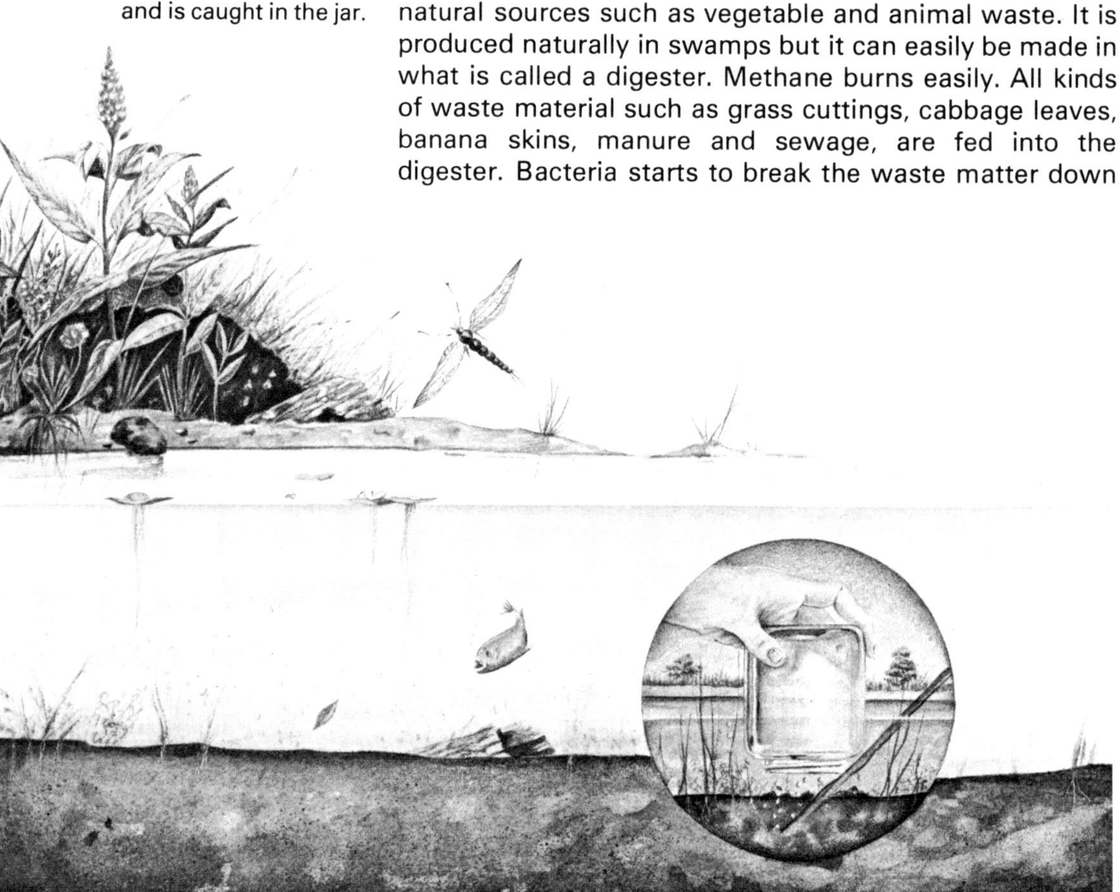

into smaller particles and, as they do so, the methane gas is produced. It is then simply led out through a pipe into a storage tank.

Methane gas has been used to operate various kinds of machinery including electric generators. It has also been used to drive motor cars and some people believe methane gas will be the motor fuel of the future. The main problems are that it takes a lot of waste to produce a small amount of gas and it takes a lot of gas to drive a car more than a few kilometres.

ENERGY PROBLEMS

The shortage of energy

The world is facing severe energy shortages in the years ahead, because we have become such big users of energy that we are using it at a greater rate each year. Every time a new gadget comes into our homes we are using more energy. Every time we use a car, we are using energy. The energy we use comes from the fossil fuels that took millions upon millions of years to form in the ground. No more will form for millions of years to come.

Coal is still one of the most important and economical sources of energy, and new seams of this fossil energy are still being discovered and exploited. It is a very versatile fuel that can be burned to produce heat and utilised by power stations to produce electricity; in addition, it can be converted in gasworks to coke.

It is not known just how much energy is still available to us in fossil fuels. There are probably large supplies of petroleum and coal under the earth's surface that we haven't yet discovered, but experts believe that there is a limit to the amount that can be found and we are using energy so quickly that it will run out quite soon. The only question is when.

Energy from the sun

In the meantime, there is nuclear energy. If all nations go ahead and use this form of energy, the energy 'crisis' will be overcome, but only for a time. Because the fuel for nuclear energy, such as uranium, is also limited, it may last less than a hundred years.

An enormous amount of research is being done into sources of energy such as solar power, wind power and wave power. The sun is also responsible for the winds and and the winds, along with the gravitational pull of the moon, cause the waves. The more we look at it, the more important the sun becomes. The energy of the sun will

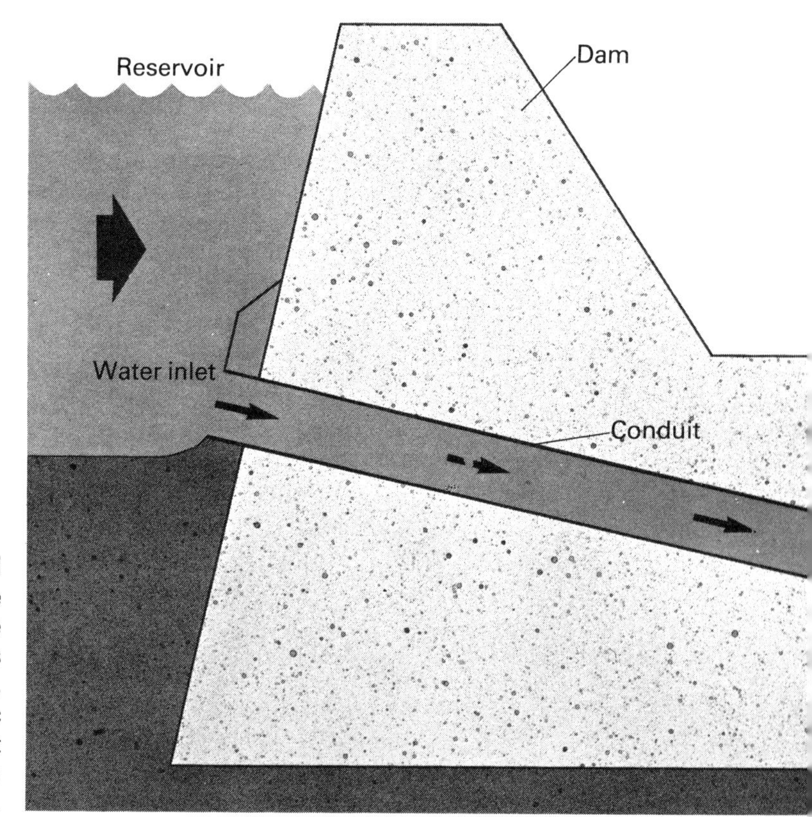

The water inlet through the wall of the dam leads from the reservoir to a large sluice gate. The force of water is used to drive a turbine, which, in turn, powers an electric generator. Electricity is sent from the power station through a grid network to consumers.

last for millions of years, all we have to do is work out the best way of capturing that energy and putting it to work for us.

Some of the possibilities are exciting. The solar cell, for example, could prove to be the major breakthrough. It may be possible to have small solar cells which are so efficient in converting the sun's rays into energy that we could run all the gadgets in our homes from a unit no bigger than a desk. On the other hand, we may be getting our electricity from giant solar furnaces or wind generators.

Hydroelectric power

Many parts of the world, including Australia, generate electricity by means of hydroelectric power. Here, the

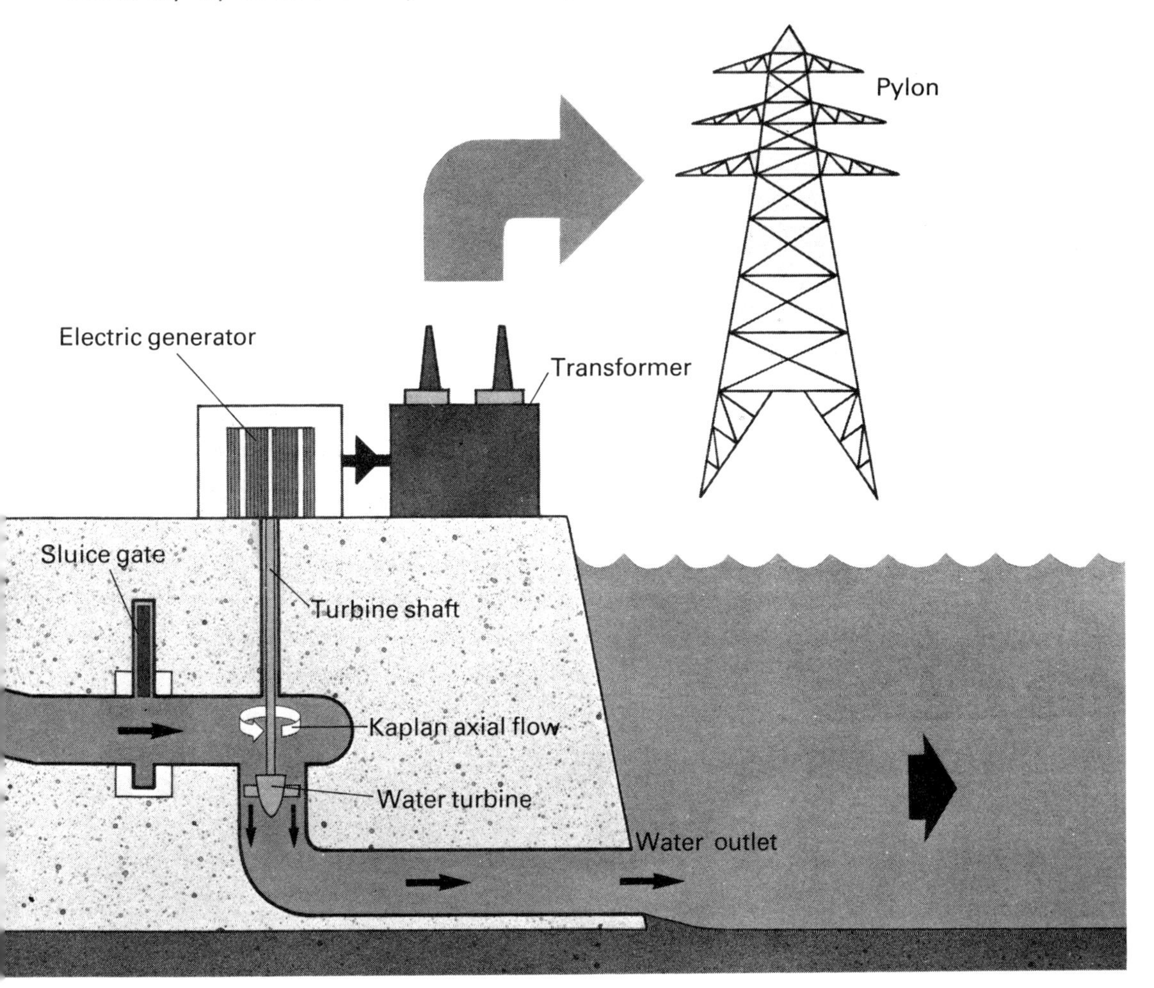

Opposite: At Rotorua, in New Zealand, geothermal power has been harnessed to produce electricity. Here, the underground heat flow has raised the temperature of the water to over 200°C, which bursts out of the earth as hot steam.

energy of falling water is used to drive electric generators, but the possibilities for increased energy production from hydroelectric power are limited. There are two reasons for this. One is that large amounts of falling water are restricted to certain parts of the world. In Australia, for example, they are restricted to Tasmania and the Snowy Mountains. The other drawback is that in parts of the world where there are huge waterfalls, they are too remote from populated areas to be economically worthwhile.

Geothermal power

Geothermal power stations use the heat that pours out of cracks in the earth's crust. Some of these are located in New Zealand where they are called *thermal springs.* They are like miniature volcanoes, but instead of erupting lava, they eject hot steam, which can be used to drive an electric generator. But thermal springs do not occur

Below: This geothermal steam station is at Wairakei, in New Zealand. In some parts of the earth's crust, the heat flow is very high and can be tapped as an alternative energy source. The hot thermal springs of New Zealand have been exploited commercially at this station in order to drive a generator.

W